Priscila do Nascimento

Assessing sustainability on construction sites

Priscila do Nascimento

Assessing sustainability on construction sites

A study in Greater Vitória - ES

ScienciaScripts

Imprint

Any brand names and product names mentioned in this book are subject to trademark, brand or patent protection and are trademarks or registered trademarks of their respective holders. The use of brand names, product names, common names, trade names, product descriptions etc. even without a particular marking in this work is in no way to be construed to mean that such names may be regarded as unrestricted in respect of trademark and brand protection legislation and could thus be used by anyone.

Cover image: www.ingimage.com

This book is a translation from the original published under ISBN 978-613-9-70593-1.

Publisher:
Sciencia Scripts
is a trademark of
Dodo Books Indian Ocean Ltd. and OmniScriptum S.R.L publishing group

120 High Road, East Finchley, London, N2 9ED, United Kingdom
Str. Armeneasca 28/1, office 1, Chisinau MD-2012, Republic of Moldova, Europe
Printed at: see last page
ISBN: 978-620-6-58270-0

Copyright © Priscila do Nascimento
Copyright © 2023 Dodo Books Indian Ocean Ltd. and OmniScriptum S.R.L publishing group

To God, my family and my husband, because without them none of this would be possible.

ACKNOWLEDGMENTS

First of all, to God, who has allowed all this to happen throughout my life, not only in these years as a university student, but who at all times is the greatest teacher anyone can know.

To my husband, my partner and best friend, who has always been attentive to my every need, always understanding and my greatest encourager in the most difficult times. Thank you for your love and affection during my journey.

To my mother, sisters and family who encouraged me and understood the moments of absence.

To my advisor, Prof.[a] Dr.[3] Luciana Aparecida Netto de Jesus, for believing in my work and for her dedication, patience and commitment during the preparation of this work.

To the teachers of the Civil Engineering course, for the harmonious coexistence, the exchange of knowledge and experiences that have been so important in my academic life.

The companies that made it possible to carry out field research and record the situations observed on construction sites.

I would like to thank everyone who, directly or indirectly, played a part in my education.

SUMMARY

The construction sector is known for its environmental impact, consuming large quantities of the planet's finite natural resources. In order to change this paradigm, a set of actions is needed linked to the introduction of sustainable practices that seek to reduce waste rates, improve quality in processes and qualify the workforce. In order to identify how sustainable a construction site can be, this paper proposes a methodology for assessing sustainability based on good practices adopted by existing environmental certification seals and research into reference cases, using a questionnaire with five themes: waste generation, water consumption, energy consumption, relationship with the environment and pollution prevention. As a case study, an assessment is presented of how the concept of sustainability is being absorbed on the construction sites of fifteen construction companies in Greater Vitória - ES, thus providing an overview of the practices adopted on the respective construction sites. In the final stage, the results of the visits were compiled and, based on the data analyzed, it was possible to ascertain the level of sustainable practices adopted during the execution of the works.

Keywords: Construction; Environment; Sustainable practices; Construction sites.</p>

TABLE OF CONTENTS

1 INTRODUCTION

In the current climate, the civil construction sector is a major player in the country's economy, characterized by the fact that it employs a large number of people directly, as well as the amount of physical and financial resources involved.

Despite these positive points, Vilhena (2007) points out that activities related to the construction, operation and demolition of buildings promote environmental degradation, resulting from the excessive consumption of natural resources and the generation of waste.

In this sense, it is also important to note the waste of materials during the construction process. According to John (2000), construction wastes an average of 56% of cement, 44% of sand, 30% of plaster, 27% of conductors and 15% of PVC pipes and conduits.

Many environmental impacts are caused during the construction process, including water and energy consumption, effluents from activities inherent to construction sites, such as water from washing concrete mixers and toilets, noise pollution, soil contamination, dust generation and air pollution.

As a result, it can be seen that the production practices adopted today by the majority of construction companies need to be adjusted so that the sector can cooperate with the country's more sustainable development.

Blumenschein (2004) explains that, considering the size and importance of the environmental impacts they cause, the construction industry can and should contribute to the search for sustainable development. Where traditionally only the tripod of time, cost and quality has been considered, environmental aspects must also be taken into account.

Over the years, a number of certification systems have emerged with the aim of encouraging actions to reduce the environmental impact of construction, such as the Building Research Establishment Environmental Assessment Method (BREEAM), Leadership in Energy and Environmental Design (LEED), Comprehensive Assessment System for Building Environmental Efficiency (CASBEE), Haute Qualité Environnementale (HQE), Green Building Tool (GBTOOL), and the Brazilian seal of High Environmental Quality - AQUA.

The aim of these building certifications is to attest to the quality and environmental performance of buildings, rather than certifying the construction company itself. Degani (2003) points out that by adopting practices per project, companies are able to induce improvements in the environmental performance of their buildings as a whole.

Even though there are important national initiatives, in terms of laws and resolutions, to adhere to environmental initiatives in the construction industry, such as Resolution No. 307 of the National Environment Council (CONAMA), which provides for waste management, it should be noted that in Brazil, in terms of legislation and public policies, Degani (2009) states that the incentives are still not sufficient to direct the construction sector towards adherence to more sustainable practices.

Dorsthorst and Hendriks (2000) consider that the challenge of reconciling environmental preservation with economic growth can only be achieved with the support of all segments of society.

The authors believe that isolated actions are not the solution to mitigating the effects of industry on the environment, nor is controlling the production cycle in such a way as to minimize the input of raw materials and the output of waste into the environment.

In this sense, the development of this research, among the subjects that will be addressed, aims to contextualize the best aspects of sustainability applied to the construction site, deepen the knowledge about managerial actions and good practices existing on construction sites.

1.1 OBJECTIVES

1.1.1 General Objective

Evaluate and classify sustainable practices on construction sites in the greater Vitória area - ES.

1.1.2 Specific objectives

- Study the operation and organization of construction sites;
- Analyze the impacts generated on the construction site;
- Identify good sustainable practices on construction sites.
- Point out measures applied to construction sites included in sustainability assessment tools.
- Draw up a questionnaire based on sustainability assessment tools, research on the subject and reference cases.
- Evaluate and classify sustainability on construction sites in greater Vitória - ES using the questionnaire developed.

1.2 BACKGROUND

Construction is one of the human activities that most affects the environment, being responsible for the consumption of 40% of natural resources, 34% of water consumption and 55% of wood consumption. Credídio (2008) points out that, in addition to the rational use of these natural resources, the sector also has to worry about the waste generated by construction work, which is then dumped in unsuitable areas, polluting cities. No less than 67% of the total mass of solid urban waste comes from construction sites.

The waste generated on construction sites stands out both for the quantity it represents and for the impacts it causes, especially when it is taken to inappropriate places. For this reason, the issue is dealt with by a federal resolution under No. 307/2002 of CONAMA, which provides for its management.

To this end, good management within the construction site is indispensable, even if not enough techniques are applied to reduce the environmental impact generated by the construction industry.

In this sense, sustainability certification tools provide knowledge for those wishing to implement sustainable certification seals in their buildings, as they provide a basis for practices to be carried out throughout the construction process.

One of the basic principles for making a building more sustainable includes: reducing waste generation, reducing water consumption, reducing energy consumption, taking advantage of local natural conditions, implementing and analyzing the surroundings, recycling and reusing solid waste.

This shows the importance of studying good sustainability practices on construction sites in order to help the construction industry develop in a correct and more sustainable way.

1.3 WORK STRUCTURE

The work is divided into three main chapters, the first of which is the introduction, which gives a general overview of the topic based on the literature review and seeks to demonstrate its relevance. It also contains the justification for the topic and describes the objectives to be achieved in the overall development of the work.

In the second chapter, the theoretical framework will be presented, describing the general concepts related to the topic of sustainability on the construction site, defining sustainable construction, presenting some of the tools for assessing sustainability and identifying the criteria related to the construction site, presenting the definition of the construction site and the importance of its organization. Some of the environmental impacts generated on construction sites will be mentioned and some examples of good practices that can be applied to construction sites will be presented.

Chapter three refers to the methodology which presents the method proposed for carrying out the research, defining the type of research, the universe and the sample, which will be collected and how the data will be processed.

Chapter four presents the case study and the results of the evaluation carried out in the field research and presents the classification of construction sites in Greater Vitória - ES.

Finally, the fifth chapter presents the conclusion of the work carried out.

2 THEORETICAL FRAMEWORK

2.1 SUSTAINABLE CONSTRUCTION

According to Pinheiro (2003), sustainability has been defined and sought after over time, although the assumption of its importance has been gaining prominence in international terms, especially in the 90s.

One of the most common definitions of sustainability is that of Brundtland (apud PINHEIRO, 2003), which is "ensuring sufficient resources for future generations to have a quality of life similar to our own".

According to Pinheiro (2003, p. 2), sustainable construction refers to:

> [...] the application of sustainability to construction activities, defined as the creation and responsible management of the built environment, based on ecological principles and the efficient use of resources.

In sustainable construction there are various terminologies, such as: Eco-Architecture, Bioarchitecture, Bioclimatic Architecture, Green Building, Natural Building, Alternative Architecture, Green *Building,* among others.

According to Schwarz (apud GEHLEN, 2008), these terms are related, but they don't mean the same thing. Each one was coined to represent the characteristics of a certain project within a specific framework.

According to Pires (2008), sustainable construction aims to offer future generations the same opportunities as today in relation to the environment and everything it provides.

The author considers that sustainable construction is a modern environmental strategy, based on reducing pollution, saving energy and water, and reducing the consumption of natural raw materials, as well as improving health and safety conditions for workers, end users and the community in general.

According to Araújo (2009), sustainable construction is a holistic process whose aim is to restore and maintain harmony between the natural and built environments, and to create establishments that affirm human dignity and encourage economic equality.

For Silva (2003), seeking a more sustainable construction industry means providing more value, polluting less, helping to make sustainable use of resources, responding more effectively to stakeholders, and improving the quality of life today without compromising the future.

Tavares (2009) points out that the concern with sustainable development gained momentum after Eco

92, when the government, companies and civil society began to work on designing instruments that took into account the social and environmental dimensions, gaining greater momentum in 2000.

The author also points out that eco-efficiency is one of the practices of sustainability, through which companies seek to produce more with less use of energy, water and materials, either by improving the efficiency of existing production processes or through innovative approaches to design, materials, equipment and construction. In this sense, he also highlighted innovation in building projects aimed at reducing energy consumption, reusing and saving water, selective collection and recycling of construction and demolition waste, among others.

Rovers (apud GEHLEN, 2008) discusses sustainability in construction on three levels: environmentally conscious construction, sustainable construction and sustainable living.

The author considers that these are the three main points that need to be managed in order to reduce the environmental impacts and climate change directly related to construction activities:

- The first, environmentally conscious buildings, is considered from the point of view of the building itself: reducing the impact of the use of energy, water and material resources, including waste.
- The second level: sustainable buildings, considers all aspects related to buildings and the environment, i.e. flora, fauna, infrastructure, air quality and urban design.
- The third level, healthy living, considers the population's daily way of life, in a way that guarantees a high standard of living and means that policies and economic actions work together to increase general well-being.

For Agopyan and John (2012), the initial milestone for sustainable construction took place in 2000, with the International Congress of the International Council for Research and Innovation in Building and Construction (CIB) - Symposium on Construction and Environment - theory into practice.

The author reports that at this event, proposals for sustainability were presented, which even contributed to Agenda 21 on sustainable construction for developing countries:

- Reduction of material losses in construction;
- Increased recycling of waste as building materials;
- Energy efficiency in buildings;
- Water conservation;
- Improved indoor air quality;
- Durability and maintenance;
- Reducing the housing, infrastructure and sanitation deficit;

- Improving the quality of the construction process;

The following are tools that evaluate these measures in an integrated way.

2.2 SUSTAINABILITY ASSESSMENT CRITERIA FROM THE POINT OF VIEW OF THE CONSTRUCTION SITE

According to Degani (2009), many companies in the construction industry are currently increasingly concerned about achieving a more correct environmental performance by controlling the impacts of their activities and products and services on the environment.

Lobo (2010) points out that, through private and voluntary initiatives, programs have been created in various countries to assess and certify sustainable built environments. These assessment methods are increasingly being disseminated and introduced into the construction sector, as they bring benefits such as the increased value of the certified project, among others.

It is clear that the certification seals have been proposed to make the construction process less harmful to the environment and society. It is therefore important to know the main sustainability items indicated in each environmental certification system.

In this study, which aims to develop a questionnaire to assess sustainable practices on construction sites, four assessment tools were chosen as a basis. The choice of these tools is based on the selection of standards and tools that are currently available and more widespread in the Brazilian market. These are: ISO 14001, LEED, AQUA and the Blue House Seal.

2.2.1 NBR ISO 14001

According to the ISO 14000 series of standards (2004), defined as a set of non-mandatory environmental standards with an international scope, companies can obtain environmental certification if they have an Environmental Management System (EMS) in place for their production processes. With an EMS in place, companies start to encourage recycling, look for less impacting raw materials and processes, and rationalize the use of renewable and non-renewable natural resources.

NBR ISO 14001 (2004, p. 1), which was drawn up by the Brazilian Environmental Management Committee (ABNT/CB-38) by the Environmental Management Study Commission (CE- 38:001.01), aims to "provide organizations with elements of an effective EMS that can be integrated with other management requirements and help them achieve their environmental and economic objectives".

This standard is based on the methodology known as Plan-Do-Check-Act (PDCA). PDCA can be briefly described as follows:

- Planning: Establishing the objectives and processes needed to achieve results in line with the organization's environmental policy;

- Execute: Implement the processes;

- Verify: Monitor and measure processes in accordance with environmental policy, objectives, targets, legal requirements and others, and report the results;

- - Act: Act to continuously improve the performance of the environmental management system.

The PDCA model is illustrated in Figure 1.

Figure 1 - Environmental management system model.

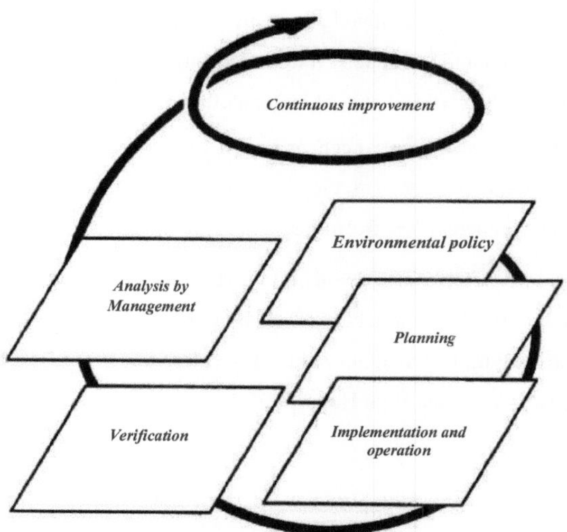

Fonte: NBR ISO 14001 (2004).

According to ISO 14001 (2004), the adoption of an environmental management system entails benefits that include: reduced use of raw materials and energy consumption; improved process efficiency; reduced waste generation and disposal costs; improved waste management, using processes such as recycling and incineration to treat solid waste, the use of more efficient techniques to treat liquid effluents and the reduction of environmental liabilities.

According to Pombo and Magrini (2008), only 2% of the total ISO 14001 certifications issued in Brazil are for construction companies, indicating a possible difficulty for the sector in adapting to environmental issues.

2.2.2 Leadership in Energy and Environmental Design - LEED certification

The Green Building Council Brazil (2015) reports that in 2000 it created the Leadership in Energy and Environmental Design (LEED), which, after being widely disseminated in the USA, influenced the creation of other systems around the world.

The author also states that LEED certification for building operations and maintenance deals with various project development processes that exist in the construction market in the USA and worldwide. These processes encompass nine rating systems for the design, construction and operation of buildings, such as:

- LEED New Constructions (NC), for new buildings or major renovation projects;

- LEED for Neighborhood Development (ND), for neighborhood development projects;

- LEED Core & Shell (CS), for projects in the envelope and central part of the building;

- LEED Retail NC and CI, for retail stores;

- LEED Healthcare, for healthcare facilities;

- LEED Existing Buildings - Operation and Maintenance (EB_OM), for maintenance projects in existing buildings;

- LEED for Schools, for schools;

- LEED for Commercial Interiors (CI), for interior projects or commercial buildings;

- LEED for Homes, for efficient residential projects.

Although there is no version for construction sites, these classification systems propose specific requirements to be considered during construction.

However, this study will focus on the LEED New Constructions (NC) rating system, which contains six criteria aimed at construction, namely: pollution prevention in construction activities, alternative transportation, construction waste management, air quality management plan during construction, regional priorities and urban density and connection with the community.

The Green Building Council Brazil (2015) points out that, among the possibilities for projects, LEED is structured on the basis of scores, varying on a scale from 0 to 100, and this scale defines the degree of certification that will be obtained. Table 1 summarizes the scores for each dimension of LEED New Construction and Major Renovations.

Table 1 - Summary of the scores for each dimension of LEED New Construction and Major Renovations.

CATEGORY	PREREQUISITES	POSSIBLE POINTS
SUSTAINABILITY OF SPACE	1	26
RATIONALIZING WATER USE	1	10
ENERGY EFFICIENCY	3	35
INTERNAL ENVIRONMENTAL QUALITY	2	15
MATERIALS AND RESOURCES	1	14
INNOVATION AND DESIGN PROCESSES	0	6
REGIONAL CREDITS	0	4
TOTAL	8	110

Source: USGBC (2015).

According to the Green Building Council Brazil (2015), there are four levels of certification that depend on the amount of points achieved: basic LEED certification (LEED Certified), silver certification (LEED Silver), gold certification (LEED Gold) and maximum platinum certification (LEED Platinum), as shown in Graph 1.

Graph 1 - LEED certification levels.

Source: USGBC (2015).

According to Oliveira (2011), LEED has some assessments related to the construction site, which require the existence of a wheel-washing system so as not to pollute the surroundings and the city; the adoption of plastic tarpaulins on the trucks so as not to drop objects and not be directly exposed to the weather; construction control through checklists, photographic surveys and lectures; preservation of native vegetation and replanting in open areas, as well as employee training in the selective collection of waste generated.

The author says that the construction site project can also include a green area for planting gardens, an erosion and sedimentation control plan.

2.2.3 High Environmental Quality Certification - AQUA

According to the Vanzolini Foundation (accessed on: October 10, 2015), the AQUA Process (High Environmental Quality), based on the French certification Démarche HQE (Haute Qualité Environmentale) was adapted to Brazilian needs by a partnership between the Vanzolini Foundation, USP's Polytechnic School and the Centre Scientifique et Technique du Batiment (CSTB).

Cardoso (2011) explains that AQUA is a project management process aimed at controlling the impacts of a new or refurbished building on the external environment, taking into account the comfort and health of users, while also ensuring the operational processes related to the planning, design and construction phases, in order to achieve the environmental quality of the building.

The author also states that this process is considered to be the first label to take Brazil's specificities into account when drawing up its 14 categories, as can be seen in Table 2.

Chart 2 - Categories of the AQUA process.

MANAGING IMPACTS ON THE EXTERNAL ENVIRONMENT		CREATE A HEALTHY AND COMFORTABLE INTERIOR SPACE	
Site and Building		**Comfort**	
1	The building's relationship with its surroundings	8	Hydrothermal comfort
2	Choose from a wide range of products, construction systems and processes.	9	Acoustic comfort
3	Construction site with low environmental impact	10	Visual comfort
		11	Olfactory comfort
Management		**Health**	
4	Energy management	12	Sanitary quality of the environment
5	Water management	13	Sanitary air quality
6	Management of waste from the use and operation of the building	14	Sanitary quality of water
7	Maintenance-Permanence environmental performance		

Source: Cardoso (2011)

The Vanzolini Foundation (accessed on: October 10, 2015) explains that for an enterprise to be AQUA certified, the entrepreneur must achieve at least one performance profile with 3 categories at the best practices level, 4 categories at the good practices level and 7 categories at the base level, as illustrated in Graph 2.

Graph 2 - Minimum performance profile for AQUA certification.

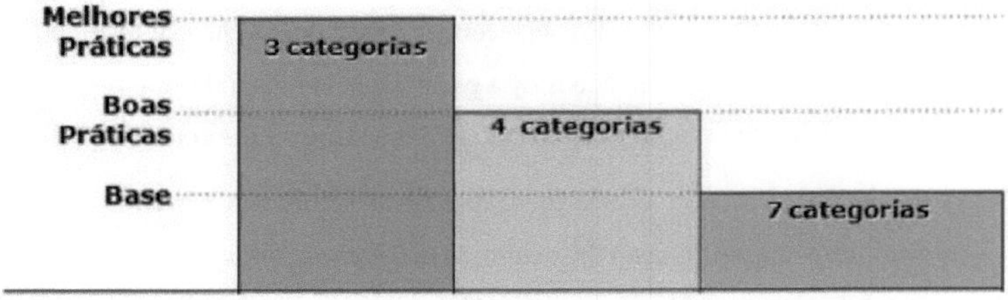

Source: Vanzolini Foundation (accessed on: October 10, 2015).

For the Vanzolini Foundation (accessed on: October 10, 2015) there are five possible classifications depending on the result obtained in each of the categories (Chart 3).

Table 3 - AQUA classifications.

Global Level	Minimum levels to be achieved
AQUA Passes	14 categories in B
AQUA Good	Between 1 and 4 stars
AQUA Very good	Between 5 and 8 stars
AQUA Excellent	Enter Sei! stars
AQUA Exceptional	12 stars or more

Source: Vanzolini Foundation (accessed on: October 10, 2015).

Oliveira (2011) considers those that fall into category 3, construction sites with low environmental impact, to be relevant for construction sites:

- Basic provisions and requirements for achieving a construction site with low environmental impact;
- Limiting nuisances;
- Limiting pollution risks that could affect the land, workers and the neighborhood;
- Construction site waste management;
- Control of water and energy resources;
- Proposal of a balance sheet with the aim of measuring the efforts and effects of the environmental provisions implemented, showing the set of elements situated at the Superior or Excellent level.

2.2.4 BLUE HOUSE SEAL certification

Caixa Econômica Federal's Casa Azul CAIXA Seal is the first project sustainability rating system offered in Brazil, developed for the reality of Brazilian housing construction. In order to improve the use of natural resources and social benefits, this seal uses realities that are appropriate to the local reality.

John and Prado (2010) report that the Seal's methodology was developed by a technical team from Caixa Econômica Federal with a wealth of experience in housing projects and sustainability management. A multidisciplinary group of professors from the Polytechnic School of the University of São Paulo, the Federal University of Santa Catarina and the State University of Campinas, which was part of a research network funded by Finep/Habitare1 and Caixa Econômica Federal, acted as consultants, even organizing a workshop which also included the participation of entities representing the market.

The authors also clarify that the CAIXA Casa Azul seal is a socio-environmental classification tool for housing projects, which seeks to recognize projects that adopt more efficient solutions applied to the construction, use, occupation and maintenance of buildings, with the aim of encouraging the rational use of natural resources and improving the quality of housing and its surroundings.

According to John and Prado (2010), with the Blue House Seal, Caixa Econômica Federal intends to establish a partnership with project proponents, providing guidelines to encourage the production of more sustainable housing.

The authors state that the Blue House Seal defines six categories:

- Category 1 : Urban quality;
- Category 2 : Design and comfort;
- Category 3 : Energy efficiency;
- Category 4: Conservation of resources and materials;
- Category 5 : Water management;
- Category 6 : Social practices.

Within these 6 categories there are 53 groups, as shown in Table 12 (Appendix A).

The Blue House Seal classifications are divided into silver, bronze and gold (Figure 2), where the minimum criteria for obtaining it are illustrated in Table 4.

Figure 2 - Casa Azul Seal logos - Gold, Silver and Bronze levels.

Source: John and Prado (2010).

Chart 4 - Blue House Seal gradation levels.

Gradation	Tremendous service
BRONZE	Mandatory criteria
SILVER	Compulsory criteria and 6 more free choice criteria
OR IRO	Compulsory criteria and 12 more Irvre choice criteria

Source: John and Prado (2010).

Some of these mandatory criteria are relevant in the construction site phase and the main one is based on CONAMA's Federal Resolution No. 307/2002, which obliges the waste generator, the construction company, to draw up a Construction Waste Management project, rated in category four - conservation of resources and materials, with the aim of establishing the necessary procedures for the environmentally appropriate handling and disposal of waste generated on construction sites.

For a better understanding of the items related to the construction phase addressed by the different sustainability assessment tools presented, the approach related to each tool is presented in Table 11 (Appendix A).

As far as ISO 14001 (2004) is concerned, it assesses whether or not good practices have been complied with, regardless of their performance, and does not specify criteria for sustainable practices applicable to construction sites like the assessment methods mentioned. For this reason, it will not be shown in the table.

2.3 ORGANIZING THE CONSTRUCTION SITE

According to Mourão, Novaes and Kemmer (2009), the time and space difficulties involved in moving around and locating facilities on construction sites are one of the biggest causes of productivity losses. According to the authors, these situations can be avoided and managed if they are identified in good time. However, companies do not always use management tools that take into account the need to allocate space on the construction site.

According to Cesar et al. (2011) it is now necessary to plan the location of each area, including work areas, storage areas and circulation areas. However, the definition of storage spaces is most often done in real time, when the materials are received.

The authors also state that this situation often results in chaotic construction sites and inefficient material handling operations. Advance knowledge of the physical flow between the various sectors is essential for the efficiency of the process as a whole.

Cesar et al. (2011) explains that the aim of planning the layout of a construction site is to make the best use of the space available for the work, locating materials, equipment and labor, so that the right conditions are created to carry out tasks efficiently, by changing the sequencing of activities, reducing distances and travel times and better preparing work stations.

There are also a number of technical and legal recommendations that need to be followed when planning a construction site. These include NR (Regulatory Standard) 18, which refers to worker health and safety conditions.

This standard lays down requirements, such as the provision and sizing of living areas and the restriction of circulation in loading areas for cranes, for example. However, compliance with legal requirements alone does not guarantee that the construction site is well organized logistically.

In this sense, it is important to note that the Ministry of Labor and Employment (MTE), through standard NR18 (2015, p. 52) called Conditions and Environment of Work in the Construction Industry, defines a construction site as "Fixed and temporary work area where support operations and the execution of a work are carried out". For NBR 12284 (1991, p 1), called Living areas on construction sites, the construction site is defined as "A group of areas intended for the execution and support of work in the construction industry, divided into operational areas and living areas".

Cesar et al. (2011) points out that on a construction site, the focus should be on minimizing interference between production sectors, living areas and administrative areas, prioritizing quality of life at work along with productivity and quality of services, and evaluating the system's capacity to meet production and the cost of alternatives.

It is therefore important to note that a sustainable construction site is one in which waste, improvisation, accidents, environmental impacts and nuisances to the neighborhood and surroundings are reduced as much as possible. Therefore, with good planning, it is possible to minimize problems and reduce the generation of waste, dust, noise, traffic problems, accidents at work, pollution in general and the waste of natural resources. The impacts are felt by the construction workers themselves, neighbors, pedestrians and visitors.

2.4 ENVIRONMENTAL IMPACTS GENERATED AT THE CONSTRUCTION SITE

According to NBR ISO 14001 (2004, p.2), environmental impact can be defined as "any modification of the environment, whether adverse or beneficial, that results, in whole or in part, from the activities, products or services of an organization".

Spadotto's concept (apud Oliveira, 2011) defines environmental impact as any alteration to the physical, chemical or biological properties of the environment, caused by any form of matter or energy resulting from human activities, which directly or indirectly affect it:

I. The health, safety and well-being of the population;

II.   Social and economic activities;

III.   The aesthetic and sanitary conditions of the environment;

IV.   The quality of environmental resources.

This research will address some of the various environmental impacts, which will be conceptualized and related to measures to be applied in the construction industry.

2.4.1 Waste generation

Inojosa (2010) teaches that the study of solid construction and demolition waste - SSCD - known as "rubble", is justified by four factors: the large volume generated, the difficulty of proper final disposal, environmental risks from disposal in clandestine deposits and the potential for recycling waste.

The main public instrument regulating the management of this waste is CONAMA Resolution 307/2002, which establishes guidelines, criteria and procedures for the management of construction waste.

According to Article 2 of Resolution 307/2002, the RSCD is defined as:

> This is waste from construction, renovation, repair and demolition work, and that resulting from the preparation and excavation of land, such as: bricks, ceramic blocks, concrete in general, soils, rocks, metals, resins, glues, paints, wood and plywood, linings, mortar, plaster, roof tiles, asphalt paving, glass, plastics, pipes, electrical wiring, etc., commonly referred to as construction debris, rubble or rubble.

Figure 3 also illustrates the definition of RSCD, according to Resolution 307/2002.

Figure 3 - Origins of the RSCD.

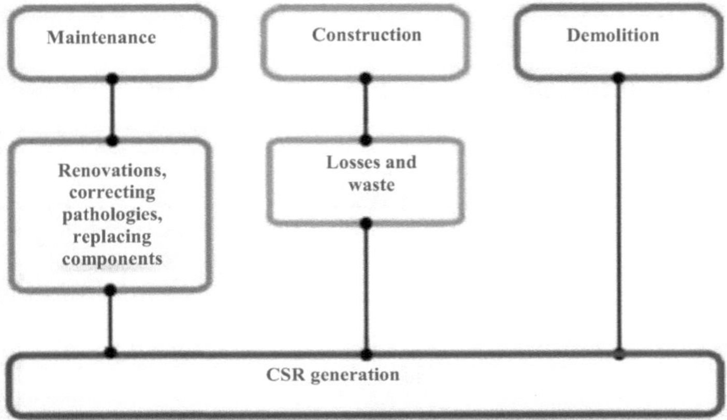

Source: Inojosa (2010).

According to Marques Neto (2005), the highest rates of CSW generation in Brazil are in construction work. The continuous and growing generation of waste in civil construction in the country is directly linked to the high level of waste of materials in the construction of projects.

According to Rocha (2006), there are some factors that cause losses in the construction industry. They are:

- Faults or omissions in the preparation of projects;
- Poor quality materials;
- Improper packaging of materials;
- Poor staff qualifications;
- Lack of appropriate techniques and equipment;
- Poor management of the construction site;
- Lack of technical support in production;
- Lack of information and guarantees regarding the sector's products and services;
- Lack of texts with procedures and systematization of knowledge;
- Lack of a culture of reuse and recycling.

Schenini, Bagnati and Cardoso (2004) explain that although there are differences between the type of work (construction or demolition), the composition of CDW is a mixture of gravel, sand, concrete, mortar, ceramic bricks and concrete blocks, wood waste, cardboard boxes, iron and plastic.

CONAMA Resolution 307/2002 classifies waste into four classes, as shown in Table 5.

Table 5 - Waste classification according to Resolution 307/2002.

CLASS	DESCRIPTION
A	Waste that can be reused or recycled as aggregates, such as concrete, mortar, ceramic components and soil from earthworks.
B	Waste for other uses, such as: wood, paper/cardboard, plastics, metals, glass, plaster, etc.
C	Waste for which no economically viable technologies or applications have been developed to enable its recycling or recovery.
D	Hazardous waste from the construction process, such as: paints, solvents, oils and others or those contaminated or harmful to health.

Source: CONAMA (2002) (adapted).

According to ABN AMRO (apud GEHLEN, 2008), the amount of construction and demolition waste generated is, on average, 150 kg/m^2 built, and construction waste makes up between 41% and 70% of the mass of solid urban waste, meaning that in many municipalities more than half of the waste generated by the entire city is construction waste.

Blumenschein (2004) explains that waste management requires prior preparation of the site and its employees. The site must have a disposal and temporary storage area sheltered from the rain, in the case of Class B and D, and easily accessible to the internal public, as well as for the removal of material from the site.

The author also states that employees should be trained and made aware of the importance of waste segregation and their responsibility in the process as a whole.

Among the advantages of adopting the practice of solid waste management by construction companies, Pinto (2005) reports that the following stand out: compliance with legal requirements; greater cleanliness of the site, contributing to greater organization of the work and a reduction in accidents at work; reduction in the consumption of natural resources, through reuse; and the consequent reduction in waste.

2.4.2 Energy consumption

According to Lamberts and Triana (2007), energy consumption has grown all over the world, mainly due to people's way of life and their growing demands for greater comfort through equipment and systems that use energy from non-renewable sources. In developing countries such as Brazil, population growth and rapid migration to cities are the main causes of increased energy consumption.

According to Abreu (2012), energy consumption in construction is significant, both in the extraction of materials and in the manufacture, transportation and processing of inputs.

However, even if the most efficient equipment possible is used, attention must be paid to the issue of consumption habits. Araújo (2009) mentions the importance of keeping machines and vehicles on the construction site only when necessary.

In addition, Araújo and Cardoso (2007) advise on the importance of rationalizing energy consumption in living areas, offices and storerooms within the construction site.

The authors explain that in living areas, systems should be installed that allow for the efficient use of energy, so that workers are responsible for using them correctly. Similarly, attention should be paid to the behavior of users in office areas and living quarters, so that lights are not left on and long baths in electric showers are avoided.

In order to reduce energy consumption and waste on construction sites, Bittencourt (2012) presented some guidelines for improvement:

- Use highly energy-efficient equipment on the construction site;
- Use compact fluorescent lamps;
- Keep machines running only when necessary;
- Carry out a strong awareness campaign at the construction site;
- Use presence sensors in places where access is infrequent, to prevent light bulbs from being left on unnecessarily;

However, you should be aware that these practices to reduce energy waste can only show positive results if there is a strong awareness campaign on the construction site, pointing out the importance and necessity of reducing consumption.

2.4.3 Water consumption

Oliveira (2011) notes that fresh water is a valuable commodity and necessary for a healthy life. However, he points out that it is a scarce commodity, which makes it necessary to preserve its quality.

Pessarelo (2008) states that in building construction, as in other types of work, water is an important element in the execution of certain services. On the construction site, the use of drinking water is related to the essential demands of its employees and is preserved in accordance with labor legislation.

The author estimates that the average daily consumption per unhoused worker is 45 liters of water per day, not including the meal, and if the meal is prepared on site, this figure would rise to 65 liters of water per day.

According to Pessarelo (2008), although water is not treated as a building material, its consumption in construction services is very high.

In this sense, Filho Neto (2013) exemplifies that an average of 160 to 200 liters of water is used to make a cubic meter of concrete, and up to 300 liters of water can be used to compact a cubic meter of landfill.

Oliveira (2011) mentions that researchers attest to the importance of water consumption in construction projects and point to the need to implement water-saving programs on building sites.

The author highlights some of these programs, such as the use of taps with automatic activation and shut-off, the installation of timers in showers to determine the time spent showering, the use of rainwater or artesian wells when possible for flushing, cleaning the site, curing concrete, dosing mortars, as well as lectures with employees to raise awareness about reducing water consumption and monthly monitoring of consumption and measures to reduce it.

For Gehlen (2008), the use of rainwater for non-potable purposes should be mandatory, even on construction sites, as most of the water consumed there is for non-potable purposes, such as cleaning and humidifying surfaces. The author points out that in addition to creating reservoirs to collect water from roofs, rainwater containment ditches can be built in the lower parts of construction sites.

2.4.4 Relationship with the environment

Pessarelo (2008) points out that during construction, the company must avoid disturbing the neighborhood by controlling noise and vibration emissions, the generation of dust and mud, or problems caused by the delivery of materials and concreting. Even though these impacts are short-lived, they cause psychological stress and health problems, as well as affecting the image of the companies involved in the construction process.

According to Guedes et al. (2011), these impacts are usually difficult to measure, and actions to minimize them should be part of the day-to-day activities of the construction site, such as containing dust and mud, avoiding work that generates noise and vibrations at night and during rest periods, thus also benefiting the internal environment for workers.

The author also points out that the occupation of the public highway, whether by dumpsters placed next to the curb or by the advancement of the construction site over the sidewalk, is also evidence of the nuisance generated for the neighborhood and can cause accidents, mainly by altering traffic on local roads.

2.4.5 Pollution Prevention

According to Resende (2007), the construction industry works by demolishing, digging, cutting, scraping and sanding and in this way contributes as a source of particulate emissions, i.e. it is a polluter.

Guedes et al. (2011) states that most construction site activities cause pollution through the emission of particulate matter (dust, smoke, fumes and mist) and other polluting gases, such as CO_2 (carbon dioxide) and SO_2 (sulphur dioxide), which are present in activities from earthmoving to finishing work.

Resende (2007) sees earthmoving and the processes of breaking, cutting and drilling, as well as improper storage and transportation of materials, as major possible sources of pollution. However, simple techniques can have a satisfactory effect on preventing pollutants and on air and soil quality during and after construction.

Resende (2007) also explains that one of the techniques is to recycle or reuse the materials used in the work and, in cases of demolition, it is important to use a physical barrier and sprinkle water during the process, minimizing the emission of dust and other polluting particles.

The author points out that transferring the earth from the backhoe to the buckets of the trucks should be done at a lower height in order to reduce the amount of dust that is dissipated into the environment. Washing truck tires before leaving the construction site (Figure 4) also prevents pollution, as the mud left behind on the streets is blown away by the winds when it dries and comes into contact with the atmosphere.

Figure 4 - Wheel washer.

Source: Martins (2010).

On construction sites, the main sources of pollutants are motor vehicles and dust.

2.5 GOOD PRACTICES ON CONSTRUCTION SITES

This topic will present examples that aim to illustrate and facilitate understanding by means of case references where some of the sustainable practices mentioned have been applied.

2.5.1 Reference Case 1: New headquarters of Lorenge SA - Vitória/ES

The current headquarters of Lorenge S.A., located in Enseada do Suá, in Vitória/ES, is a commercial building consisting of 01 floor, basement, first floor, 03 types of floors and a roof. The work was completed in January 2013, lasting 21 months. The building was certified by *LEED for New Construction and Major Renovations,* obtaining a GOLD rating.

The sustainable practices carried out during the construction phase of this development, which will be shown below, met the criteria established by the LEED evaluation method.

These sustainable practices were provided by Lorenge S.A. during the construction work and will be presented below.

2.5.1.1 Pollution

In order to reduce mud and dust, as required by the LEED assessment method, a wheel-washing system was implemented to prevent pollution of the surroundings. To this end, all vehicles had their wheels washed as they left the site, as shown in Figure 5.

Figure 5 - Wheelwashers, the area designated for cleaning vehicles.

Source: Lorenge S.A. (2015).

When using vehicles to transport materials, the company chose to buy them from suppliers closer to the project, in order to reduce pollution.

In order to control and reduce the use of tobacco smoke on the construction site, in accordance with LEED criteria, a smoking room was set up on the construction site, as shown in Figure 6.

Figure 6 - Smokehouse at the construction site.

Source: Lorenge S.A (2015).

A sedimentation and erosion control plan was carried out, in which barriers were applied around culverts or screens in culverts in order to retain sediment in rainwater runoff, as shown in Figure 7.

Figure 7 - Barrier for manhole protection.

Source: Lorenge S.A (2015).

2.5.1.2 Waste Management

According to the criteria of CONAMA resolution 307/2002 and LEED, waste management was carried out properly, separating, routing and appropriately disposing of the construction and demolition waste generated during the work, such as: concrete, mortar, ceramic components, wood, cardboard, plastic, glass, metals, plaster and others, as shown in Figure 8.

Figure 8 - Correct separation of waste.

Source: Lorenge S.A (2015).

Lorenge S.A. reported that 50% of the waste generated on site was sent to a recycling company and the rest to environmentally licensed landfills.

2.5.1.3 Reducing Water and Energy Consumption

The company also explained that an awareness program was carried out for the construction team (Figure 9), with the aim of alerting workers to the need to reduce water and energy consumption. The importance of reducing this consumption was explained, both on the construction site and in the employees' personal lives.

Figure 9 - Talk to the team.

Source: Lorenge S.A (2015).

2.5.1.4 Relationship with the environment

As can be seen in Figure 10, the organization and cleanliness of the surroundings of the construction site was maintained, and because pollution mitigation measures were carried out, as well as communication with the nearby population, the relationship with the surroundings during the work was gratifying.

Figure 10 - Site surroundings.

Source: Lorenge S.A (2015).

2.5.2 Reference Case 2: Modernization of Mineirão Stadium

GUEDES et al. (2011) explains that the proposal to modernize the Mineirão stadium (Governador Magalhães Pinto stadium) in Belo Horizonte followed FIFA's recommendations for the creation of Eco-arenas, i.e. ecological-green stadiums.

The author says that the idea was that, like Mineirão, all Brazilian stadiums should be built in such a way as to have the least environmental impact, without wasting materials and with greater energy efficiency.

Located in Belo Horizonte - MG, the work was completed in December 2012, lasting two years and was of a large size, as can be seen in Figure 11. After its completion, Mineirão became the first Brazilian stadium to receive the U.S. Green Building Council's (USGBC) Platinum seal, the highest category in Leadership in Energy and Environmental Design (LEED) certification.

Figure 11 - Mineirão Stadium under construction.

Source: Lott (2011).

According to Guedes et al. (2011) and Lott (2011), it is possible to identify sustainable practices applied to the Mineirão stadium construction site, as presented below.

2.5.2.1 Pollution Prevention

In order to prevent pollution during construction activities, a wheel-washing system was installed, as required by LEED, with the aim of avoiding the dust left behind by the streets (Figure 12). A positive point is that the water used to wash the wheels came from the rainwater collection system (Figures 13 and 14).

Figure 12 - Wheel washer system.

Source: Lott (2011).

Figure 13 - Water collection system from the rain.

Source: Lott (2011).

Figure 14 - Water collection system from the rain.

Source: Lott (2011).

As a result of the rains, to control soil erosion, grass was planted on the slopes and the culverts were protected around them with strong putty (Figures 15 and 16).

Figure 15 - Sedimentation and erosion control.

Source: Lott (2011).

Figure 16 - Sedimentation and erosion control.

Source: Lott (2011).

To improve the air quality on site, the construction
site was irrigated every two or three hours. The water used came from the rainwater collection system (Figures 17 and 18).

Figure 17 - Air quality plan.

Source: Lott (2011).

Figure 18 - Air quality plan.

Source: Lott (2011).

2.5.2.2 Waste management on site

Guedes et al (2011) reports that the waste generated in the first demolitions at the stadium, such as reinforced concrete, masonry, blocks and mortar, was used to build ramps for machinery access inside the stadium and also reused in other infrastructure works in Belo Horizonte.

The author also clarifies that the concrete resulting from the demolition of part of the stadium in the retrofit work was sent to the Belo Horizonte City Hall Recycling Plant with the aim of recycling it for the construction of sidewalks, street paving and as a base for the application of asphalt.

During the construction work, the waste separation control on the site was carried out correctly, using bays for storing the different classes of waste and waste garbage cans. To ensure correct separation, they were spread around the site to encourage everyone to separate their waste, as can be seen in Figures 19, 20 and 21.

Figure 19 - Waste sorting

Source: Lott (2011).

Figure 20 - Waste sorting.

Source: Lott (2011).

Figure 21 - Waste sorting.

Source: Lott (2011).

An analysis of the practices adopted at the Lorenge S.A. and Mineirão construction sites shows that these are simple measures aimed at reducing the impact on workers, the neighborhood, pedestrians and visitors to the site, as well as preserving the environment by controlling pollution on site, separating waste correctly and reducing water and electricity consumption.

It can be seen that the solutions to reducing waste generation lie in the development of sorting processes, selective collection, adequate storage for reuse, recycling or processing. In addition, provision must be made for organizing storage and circulation areas, taking care of site logistics, planning collections, among other various measures.

Saving energy and water also characterize a sustainable construction site, in addition to respecting the neighborhood with measures such as reducing noise from machinery and equipment, as well as the circulation of vehicles.

3   METHODOLOGY

According to Vergara (2010), methodology can be understood as the set of procedures needed to prove hypotheses. Each field of knowledge needs to draw up sets of rules and procedures that form part of its research, using the same techniques for all experiments.

The author states that this is what guarantees, with greater security and economy, the possibility of achieving the objective with valid and true knowledge, tracing the path to be followed, detecting errors and helping the scientist's decisions.

3.1   TYPE OF RESEARCH

According to Vergara (2010), research has two criteria: ends and means.

In the first instance, it can be said that this study is descriptive and exploratory in nature. The author states that exploratory research is carried out in a field of study with little accumulated and systematized knowledge. As for the descriptive character, it refers to research that exposes the characteristics of a certain population or phenomenon, in which it determines correlations between variables and defines their nature.

As for the means, there are also various classifications made by Vergara (2010), in which case the so-called field research was applied, since questionnaires (Appendix B) and technical visits to construction sites were carried out.

3.2   UNIVERSE AND SAMPLE

Vergara (2010) defines universe and sample as the entire sample population. In this sense, the population is understood not as the number of inhabitants of a place, but as a set of elements (companies, products, people, for example) that have characteristics that will be the object of study.

With regard to the universe of this research, it is represented by construction sites in Greater Vitória - ES, currently underway. The sample, defined as part of the universe, was chosen according to criteria of representativeness, which in this study was represented by 15 (fifteen) selected construction companies, with works in progress.

The aim was to reach the largest sample of construction sites by approaching the construction companies in question, in order to verify their conduct on their respective construction sites.

3.3   DATA COLLECTION

Initially, approaches were made to the selected construction companies, checking the number of works in progress, as well as their type and size.

Subsequently, a questionnaire was applied, based on the sustainability tools with criteria that refer to the construction sites presented in this research, as well as course completion works, master's dissertations on the subject and reference cases.

In this regard, it should be noted that this questionnaire is structured with questions pertinent to the subject of this research, such as: pollution, waste, water, energy and the relationship with the environment.

The questions asked refer to how waste is managed on the construction site, concerns about reducing pollution and impacts on the site's surroundings, reducing the emission of contributing gases, the correct storage of dangerous products, reducing dust when cleaning the site, ways of reducing water and energy consumption and measures for good relations with the neighborhood.

After applying the questionnaire, a photographic survey was carried out at one of the construction sites of each construction company, thus making it possible to verify the veracity of the data collected.

3.4   DATA PROCESSING

After evaluating the questionnaire with the data from the construction sites and relating it to the technical visits to the respective sites, we decided to carry out a classification of the construction sites by level of practices observed, using percentage compliance, as shown in Table 6.

Chart 6 - Classification of construction sites.

CLASSIFICATION OF CONSTRUCTION SITES		
Level	**Classification**	**Percentage met**
Level 0	Unsustainable construction site	0 % a 25 %
Level 1	Construction site with few sustainable practices	26% a 50 %
Level 2	Construction site with a good level of sustainable practices	51 % a 75 %
Level 3	Construction site with an excellent level of sustainable practices	76 % a 100%

Source: The author (2015).

In the case of this research, a minimum score of 25% of the practices was adopted, and a construction site in this range is not considered sustainable.

Next, the classification of levels was established according to the data analyzed in the field research. In this way, it was stipulated that adherence to 26% to 50% of the practices classifies the site as having few sustainable practices, with adherence of 51% to 75% it will be considered a site with a good level of sustainable practices and from 76 to 100% are the sites with an excellent level of sustainable practices.

It was noted that sustainability tools use percentage evaluation to analyze their criteria, as in the cases of AQUA and the CASA AZUL SEAL, mentioned in this research. To this end, the same form of evaluation was used in this research.

After obtaining the classification of the construction sites analyzed, it was possible to obtain a percentage assessment, making it possible to verify which site is more sustainable, as well as which theme is more or less prevalent among the construction sites studied. It was also possible to get an overview of how sustainability has been dealt with in the construction phase of buildings in the greater Vitória - ES area.

4 CASE STUDY

Visits were made to the construction sites of 15 construction companies located in greater Vitória - ES, with the aim of evaluating the sustainable practices employed by means of a questionnaire with 27 questions. The construction companies had residential projects in the superstructure and sealing phase and, in total, had 57 projects in progress.

A field visit was made to one of the construction sites of each construction company selected to verify the veracity of the data collected and to carry out a photographic survey. The data analyzed in this study comes from visits to 15 construction sites in progress.

Of the companies interviewed, only one had LEED certification for its headquarters and AQUA-HQE certification (in the pre-project phase) for one of its developments.

According to data provided by Inmetro (2016), no construction company evaluated had NBR ISO 14001 certification.

In this context, based on the study of sustainable practices, it was possible to assess sustainability in the activities carried out on construction sites.

4.1 ANALYSIS OF DATA COLLECTED ON CONSTRUCTION SITES

In order to better understand the data, the questionnaire was divided into themes (Figure 22) and then bar and radar graphs were generated with the averages of the results obtained.

Figure 22 - Themes covered during the evaluation.

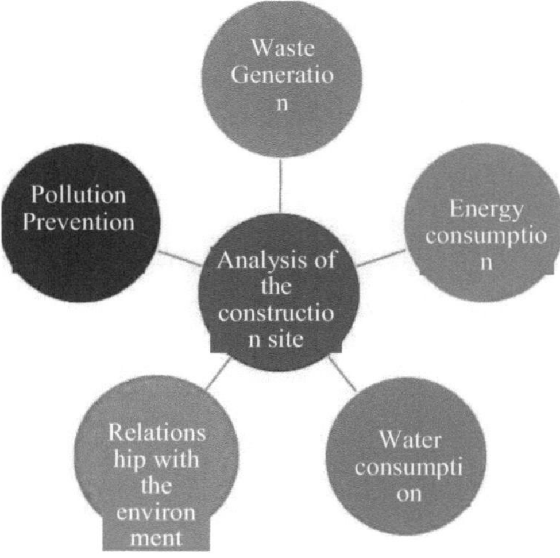

The questions for each theme were described in the graphs, with the presentation of images taken at the construction sites and the results obtained for each one.

The research concluded with a general assessment of the construction sites analyzed, by average compliance or not with the sustainable practices described in the questionnaire, making it possible to classify the construction sites by levels of observed practices, according to the percentage compliance described in Table 6.

4.1.1 Individual Analysis of Proposed Topics

4.1.1.1 Waste Generation

This topic assessed the management of waste generated on the construction site. Practices such as selective collection, loss control, the reuse of materials, as well as procedures to raise awareness among workers on site were analyzed.

By comparing the information obtained in the questionnaire with the field research, it was possible to see that the construction companies carried out good waste management on their construction sites, promoting the proper collection of waste and forwarding it appropriately by licensed carriers.

It is worth noting that the good practice common to all the construction companies evaluated is justified by the need to comply with Municipal Law 4999/2010 and CONAMA Resolution 307.

Chart 7 shows the subject of the assessment and the purpose of its application.

Table 7 - Explanatory framework for the waste management approach.

	Subject	What you should cover
1	Do you implement solid waste management on the construction site?	The construction site must separate its waste correctly according to the classes covered by CONAMA Resolution 307.
2	Is there selective waste collection on site?	Waste must be collected by class, according to its separation.
3	Are workers aware of waste management on site?	Workers should be informed about the importance of proper waste management through meetings and illustrative notices.
4	Do you reuse materials?	Address whether there is any reuse of materials that would otherwise be discarded.
5	Do you control losses on construction sites?	Address whether they carry out any loss control of the materials used.
6	Is recyclable waste disposed of correctly?	Analyze whether waste that can be recycled is sent to recycling companies.
7	Do contractors receive any guidance on how to store the debris they generate?	To find out whether the employees of contractors are instructed in the correct separation of the waste they generate.

Source: The author (2016).

Graph 3 shows the results obtained in this evaluation, in percentages.

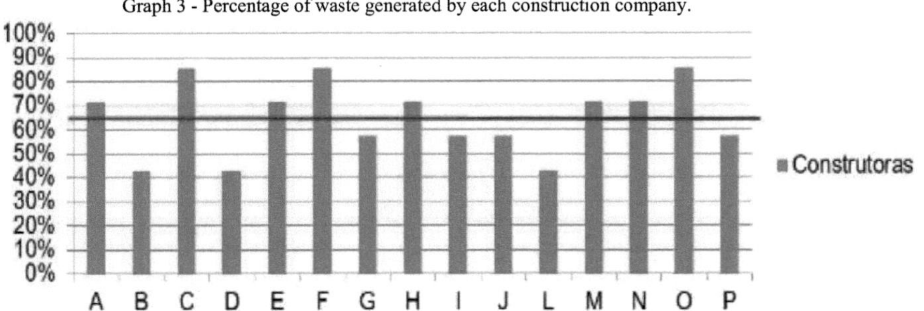
Graph 3 - Percentage of waste generated by each construction company.

It was found that construction companies were above 40%, reaching 85% in carrying out good practices in this area, reaching an average of 64.76%, as indicated by the red line (Graph 3).

Graph 4 shows the description of the questions relating to this topic, with an average of the answers given by the construction companies at the fifteen sites visited.

Graph 4 - Assessment of waste generation.

Graph 4 shows that the construction companies managed the waste generated on their sites well, with correct separation, collection and disposal, totaling 100% compliance with the sustainable practices assessed (Figures 23 to 30).

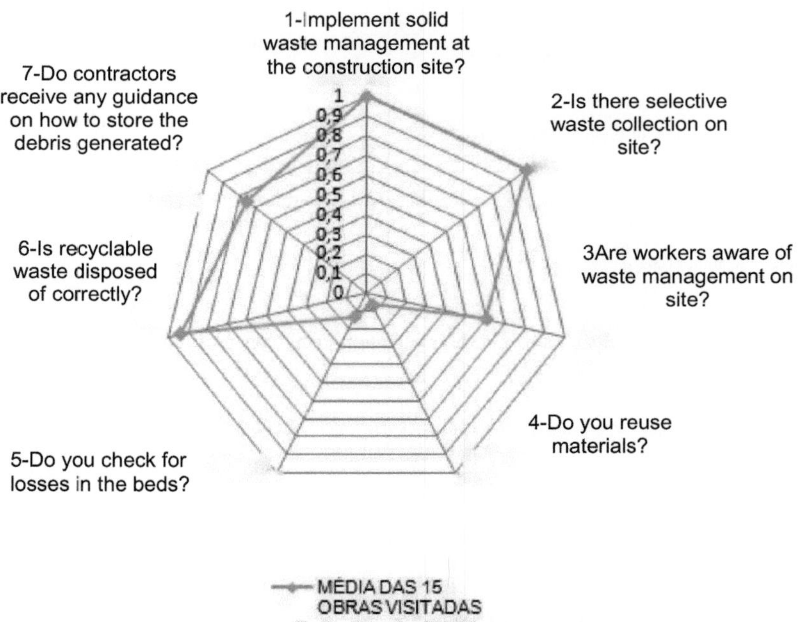

Fonte: O autor (2016).

Figure 23 - Waste separation bays.

Source: The author
(2016).

Figure 25 - Waste garbage cans with notices for
selective collection.

Source: The author
(2016).

Figure 27 - Separation of construction waste
- Class A.

Source: The author
(2016).

Figure 24 - Waste separation bays.

Source: The author
(2016).

Figure 26 - Waste garbage cans with notices for
selective collection.

Source: The author
(2016).

Figure 28 - Separation of Class A
construction waste.

Source: The author
(2016).

Figure 29 - Selective collection of wood and plastics - Class B.

Source: The author (2016).

Figure 30 - Selective collection of wood and plastics - Class B.

Source: The author (2016).

It was observed that the reuse of materials and the control of losses are not measures adopted on construction sites, measures which would imply a reduction in debris as a result of the reuse of what would otherwise be discarded.

In addition, despite the importance of the issue, 40% of the construction sites evaluated did not raise awareness among workers. Such awareness would lead to better employee performance when carrying out tasks related to waste generation.

In 75% of the construction sites, there was a concern to advise contractors on the correct management of their waste.

4.1.1.1 Energy consumption

First of all, it's important to note that energy reduction is covered by various certifications. In this way, actions aimed at reducing or eliminating energy waste are essential when it comes to sustainable practices on the construction site.

In this area, four questions were used to check the practices used to reduce electricity consumption, namely: whether priority was given to natural lighting and ventilation in the temporary facilities, whether fluorescent lights were used, whether presence sensors were installed in places where people were not present and whether awareness campaigns were implemented in routine meetings.

The results are shown in Graph 5.

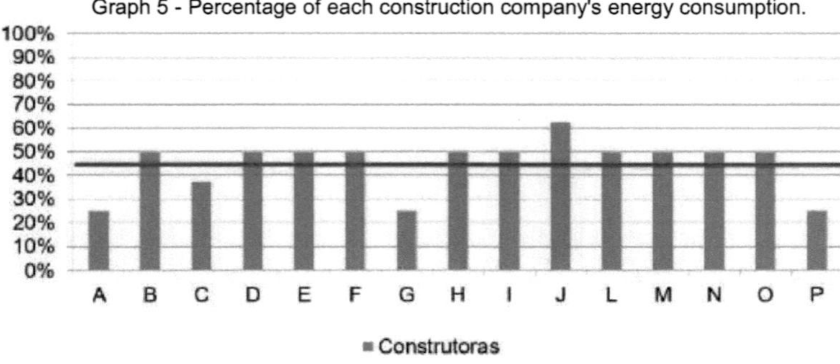

Graph 5 - Percentage of each construction company's energy consumption.

Source: The author (2016).

It was found that the construction companies obtained an unsatisfactory result, with the majority achieving a percentage of 50% in the exercise of good practices. Only construction company J exceeded 60%, resulting in an average of 45%, as shown in the red line.

Graph 6 shows the description of the questions relating to this topic, with an average of the answers given by the construction companies at the fifteen sites visited.

Graph 6 - Evaluation of energy consumption.

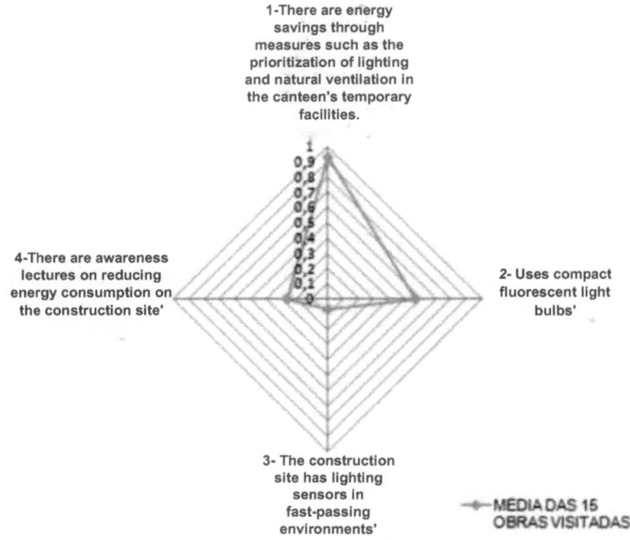

Source: The author (2016).

Graph 6 shows that construction companies take advantage of natural lighting and ventilation in their temporary facilities, avoiding the continuous use of electricity in these places, as can be seen in Figures 31 and 32.

Figure 31 - Temporary installations prioritizing natural lighting and ventilation.

Source: The author (2016).

Figure 32 - Temporary installations prioritizing natural lighting and ventilation.

Source: The author (2016).

It was also noted that the use of fluorescent lamps is not prioritized on construction sites.

Incandescent bulbs, which have high energy consumption and a short lifespan, are used in 33% of the buildings. Two types of bulb (fluorescent and incandescent) were used in 20% of the projects and 47% of them used more efficient bulbs, in this case fluorescent.

Only site "E" used LED lamps, and the site manager reported that there were savings of 60% in energy consumption.

Figures 33 to 36 show the different types of lamps used on the construction sites.

Figure 33 - Flowerbeds using incandescent bulbs.

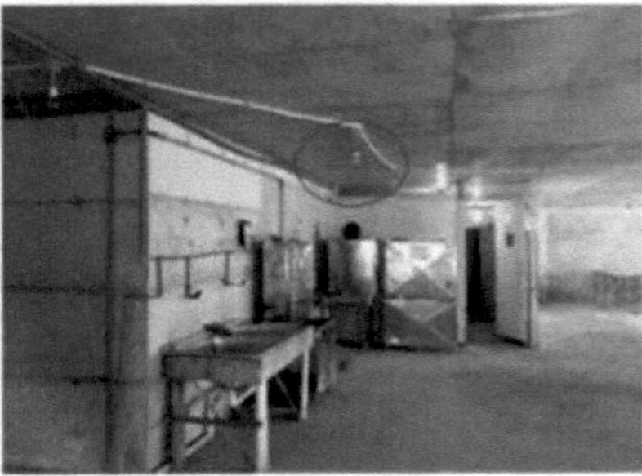

Source: The author (2016).

Figure 34 - Flowerbeds using incandescent bulbs.

Source: The author (2016).

Figure 35 - Flowerbeds using fluorescent and LED lamps.

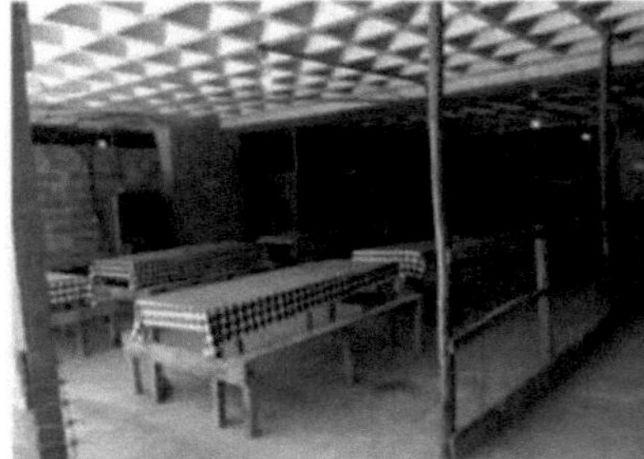

Source: The author (2016).

Figure 36 - Flowerbeds using fluorescent and LED lights.

Source: The author (2016).

Presence sensors were not used on any of the sites visited. As a result, many rooms consumed electricity unnecessarily, especially those that were little frequented.

It emerged that only one of the construction sites visited carried out awareness campaigns to reduce energy consumption.

It is worth considering that these campaigns are of the utmost importance in preventing energy waste, as well as encouraging employees to carry out these daily measures in their own homes.

4.1.1.3 Water consumption

When it comes to water consumption, especially on construction sites, waste is considered serious damage to the environment, and it is of the utmost importance to avoid waste and reduce consumption.

Following the research carried out, it was possible to see that construction companies do not show concern about the issue, and do not adequately follow the procedures for reducing water consumption.

A valid option is to capture water from wall-mounted air conditioners in temporary facilities located on construction sites, for use in tasks where drinking water is not required.

Three procedures were considered in this modality: whether the construction sites had any rainwater harvesting devices; whether artesian well water was used on the sites; and whether workers were made aware of the need to save water through talks and meetings.

The results are shown in Graph 7.

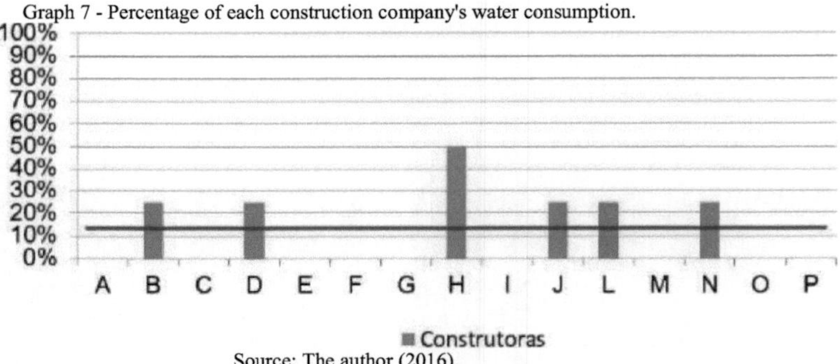
Graph 7 - Percentage of each construction company's water consumption.

Source: The author (2016).

It was found that several construction companies did not carry out any good practices in relation to water consumption, with a result of 0%. Of the construction companies visited, 60% had no procedures in place to reduce water consumption, 33% reported that they had complied with at least one of the procedures covered by this topic and only one company had carried out two of the procedures.

The average result was 11.67%, as can be seen in the red line in Graph 7.

Graph 8 shows the description of the questions relating to this topic, with an average of the answers given by the construction companies at the fifteen sites visited.

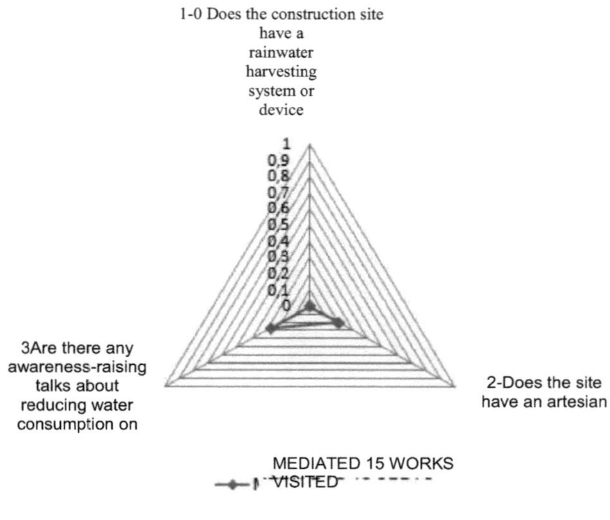

Source: The author (2016).

Graph 8 shows that none of the construction sites collected rainwater for reuse on site, as a way of saving drinking water for activities where it is not needed.

Only 20% of the works used water from an artesian well.

With regard to raising awareness among workers about saving water, only 26% of the sites held talks and meetings.

It's worth emphasizing that applying this last procedure would enable employees to use water consciously and reapply the measures in their daily lives.

A photographic survey was carried out based on the information obtained on this subject (Figures 37 to 40).

Figure 37 - Artesian well at the construction site.

Source: The author (2016).

Figure 38 - Artesian well at the construction site.

Source: The author (2016).

Figure 39 - Using water from the
artesian well for cleaning.

Source: The author (2016).

Figure 40 - Using water from the
artesian well for cleaning.

Source: The author (2016).

4.1.1.2 Pollution Prevention

Pollution prevention is an important issue because, when properly applied, it prevents health problems and reduces damage to the environment.

The dust released into the air doesn't disappear and can cause numerous health problems, from eye and skin irritation to respiratory and heart problems, as well as affecting vegetation. This is why the application of sustainable practices on construction sites is necessary to alleviate these problems, and should be concerned mainly with air, soil and water pollution.

Thus, the aim of this topic is to evaluate the application of sustainable pollution prevention practices on construction sites, where the procedures are easily accessible, making the work more efficient.

Chart 8 shows the subject of the assessment and the purpose of its application.

Table 8 - Explanatory framework for the pollution prevention approach.

	Subject	What you should cover
1	Is there public transport near the site?	There should be access to public transport close to the development in order to encourage efficient transportation.
2	Is an appropriate selection of materials and suppliers made, taking into account the reduction of environmental impacts and the emission of contributing gases?	The choice of materials must be made in such a way as to give preference to manufacturers of products with the lowest environmental impact in terms of energy consumption and the depletion of natural resources.
3	Does the construction site have bike racks?	The presence of a bicycle rack to encourage employees who live nearby to use bicycles as transportation.
4	Does the construction site have a system for washing vehicle wheels?	Use of the truck wheel cleaning device to avoid dirtying the surroundings.
5	Is smoking allowed on site?	Indoor tobacco smoke control.
6	Is the land located in an environmental preservation area?	Mitigation of ecological impacts.
7	Are contractors included in environmental education programs for employees?	Carry out environmental education with employees through lectures.
8	Are stocks of oil, fuel and other flammable products identified?	They must be protected and identified to avoid any accidents.
9	Are the rooms regularly humidified?	Cleaning with water sprays to control sources of pollution.
10	Does the site have a decanting system for the water used to wash the concrete mixer?	Construction of settling tanks to control soil and water pollution.

Source: The author (2016).

The result obtained in this evaluation can be seen in percentage terms in Graph 9.

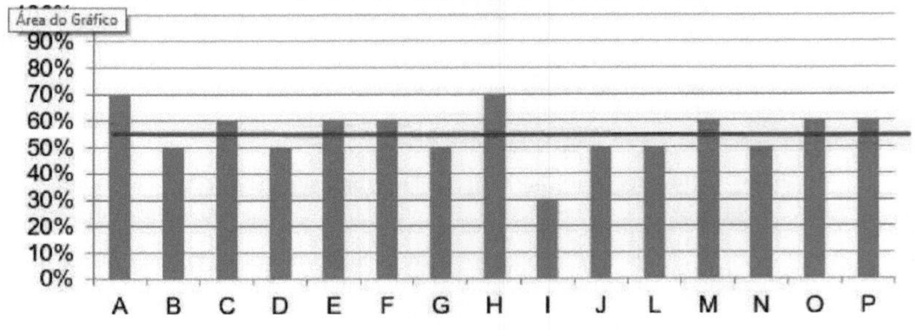

Graph 9 - Percentage of each construction company in pollution

■ Construction companies

Source: The author (2016).

It was found that several construction companies comply with between 30% and 70% of the procedures outlined in this topic, with an average of 58%, as shown by the red line in Graph 9.

Graph 10 shows the description of the questions relating to this topic, with an average of the answers given by the construction companies at the fifteen sites visited.

Graph 10 - Assessment of pollution prevention.

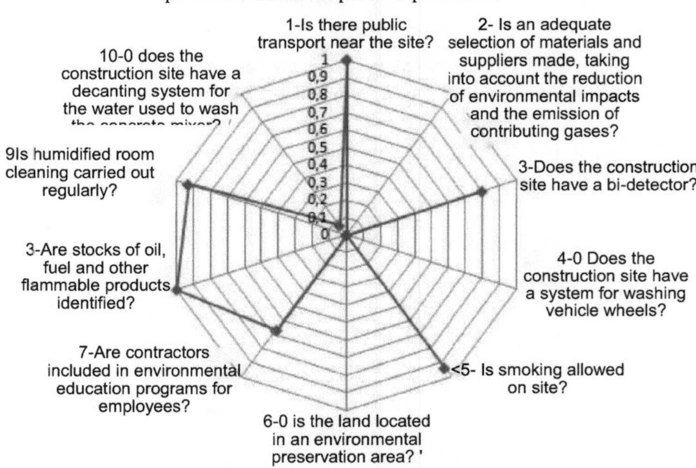

--Average of the 15 works visited

Source: The author (2016).

Graph 10 shows that all the construction sites have easy access to public transport.

As far as the purchase of materials is concerned, the result was not satisfactory, so all purchases were made at cost. Even with the small difference between the values, construction companies chose to buy materials in other states, instead of prioritizing purchases in closer locations, avoiding long journeys and the consequent reduction in CO emissions$_2$, for example.

Regarding the existence of bicycle racks, 80% of the sites have adequate infrastructure (Figures 41 and 42), encouraging employees to use this means of transportation to get to the site with a suitable place to store them.

Not all the construction sites had bike racks, as can be seen in Figures 43 and 44.

Figure 41 - Existence of bicycle racks on the construction site.

Source: The author (2016).

Figure 42 - Existence of bicycle racks on the construction site.

Source: The author (2016).

Figure 43 - No bike racks on the construction site.

Source: The author (2016).

Figure 44 - No bike racks on the construction site.

Source: The author (2016).

With regard to the wheel washing system, which consists of washing the wheels of the cars used in the construction and avoiding dirtying the streets of the neighborhood, it has not been implemented on any site.

Almost all of the construction sites allowed the use of cigarettes, except in the cafeteria, in compliance with law 12.541/2011, which prohibits smoking in partially enclosed spaces. Only one construction company completely banned smoking in the workplace. It is worth noting that the use of cigarettes affects air quality for workers and visitors.

An environmental education program was carried out on 66% of the construction sites, advising all employees on the importance of environmental preservation. It should be noted that none of the works visited were being built in an environmental preservation area.

On all sites, flammable products were stored correctly (Figures 45 and 46).

Figure 45 - Stocks of flammable products.

Figure 46 - Stocks of flammable products.

Source: The author (2016).

Source: The author (2016).

Only on site "I" was no humidified cleaning carried out in the rooms, causing dust generation and poor air quality.

On the rest of the construction sites, this cleaning was carried out frequently, as can be seen in Figures 47 and 48.

Figure 47 - Humidified cleaning.

Source: The author (2016).

Figure 48 - Humidified cleaning.

Source: The author (2016).

Despite being an excellent option for reducing water and soil pollution, only one construction site used a decanting system to wash the concrete mixer (Figures 49 and 50). This method helps to reduce the concentration of solids in the discarded water, preventing contamination of the soil and groundwater.

Figure 49 - Decanting system for concrete mixer wash water.

Source: The author (2016).

Figure 50 - Decanting system for concrete mixer wash water.

Source: The author (2016).

4.1.1.5 Relationship with the environment

This topic assessed the relationship between the construction site and its surroundings, checking that the areas around the site were clean, that sidewalks and streets were clean and paved, and that sources of pollution such as dust and noise, as well as nuisances, were avoided on these routes.

The result obtained in this evaluation can be seen in percentage terms in Graph 11.

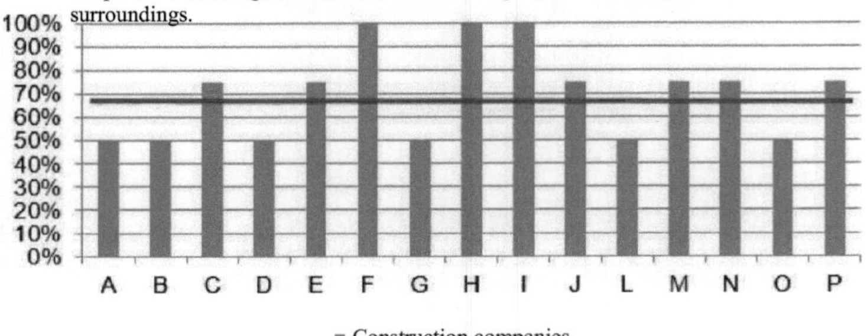

Source: The author (2016).

It was found that the construction companies were concerned about the negative impacts they could have on the surrounding area. The construction companies were above 50% in their compliance with good practices, with an average of 69.20%, as can be seen in the red line in Graph 11.

Graph 12 shows the description of the questions relating to this topic, with an average of the answers given by the construction companies at the fifteen sites visited.

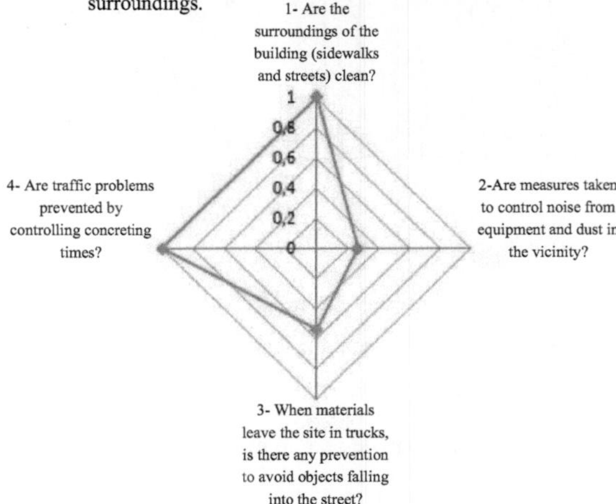

Fonte: O autor (2016).

Graph 12 shows that all the construction sites cleaned and organized the sidewalks around them (Figures 51 and 52).

Figure 51 - Sidewalks around the building.

Source: The author (2016).

Figure 52 - Sidewalks around the building.

Source: The author (2016).

Equipment noise and dust control was carried out on only 26% of the construction sites. The measures applied were: avoiding noise production before 8 a.m., using noisier equipment in locations further away from the neighborhood, when possible.

The concrete mixer (Figure 53) is a piece of equipment that generates a lot of noise and dust. However, only in two of them was there a fence around this equipment, due to complaints from neighbors. It was not possible to capture images of this enclosure.

Figure 53 - Concrete mixer on a building site.

Source: The author (2016).

Another measure adopted by the construction companies was to avoid dirt in the neighborhood as a result of the work. Figure 54 shows the installation of guards to prevent splashes on surrounding buildings during the pouring of a slab.

Figure 54 - Protection to avoid dirt in the surroundings.

Source: The author (2016).

In 53% of the construction sites, the trucks were protected with tarpaulins to prevent materials from falling onto the streets as they left the site. Figure 55 illustrates this practice.

Figure 55 - Protecting materials on a truck with a tarpaulin.

Source: The author (2016).

Sidewalks and roads used on concreting and load receiving days were signposted at all the construction sites.

This measure is applied to ensure pedestrian passage, in addition to cleaning the sites after concreting. Figure 56 shows that the signs are adequate and that pedestrian crossings have not been affected.

Figure 56 - Adequate signage on concreting and load receiving day.

Source: The author (2016).

4.2 CLASSIFICATION OF CONSTRUCTION SITES

Based on the data collected through the application of a questionnaire divided into themes and site visits, it was possible to obtain an overview of the levels of sustainable practices on construction sites.

The result obtained was calculated as the average of the five themes already described in this case study (Chart 9), namely: Waste Generation, Energy Consumption, Water Consumption, Pollution Prevention and Relationship with Surroundings.

Table 9 - Average scores for questionnaire items.

RESULT PER ITEM	1. Waste generation	2. Energy consumption	3. Water consumption	4. Pollution prevention	5. Relationship with the environment	Average per job
Work A	71%	25%	0%	70%	50%	43%
Work B	43%	50%	25%	50%	50%	44%
Work C	86%	38%	0%	60%	75%	52%
Work D	43%	50%	25%	50%	50%	44%
Work E	71%	50%	0%	60%	75%	51%
Work F	86%	50%	0%	60%	100%	59%
Work G	57%	25%	0%	50%	50%	36%
Work H	71%	50%	50%	70%	100%	68%
Work I	57%	50%	0%	30%	100%	47%
Work J	57%	63%	25%	50%	75%	54%
Work L	43%	50%	25%	50%	50%	44%
Work M	71%	50%	0%	60%	75%	51%
Work N	71%	50%	25%	50%	75%	54%
Work O	86%	50%	0%	60%	50%	49%
Work P	57%	25%	0%	60%	75%	43%
ITEM AVERAGE	65%	45%	12%	55%	70%	49%

Source: The author (2016).

The works were evaluated using a questionnaire, with the results divided into themes. The most important theme was the relationship with the surroundings and the least important was water consumption (Chart 9).

Table 10 shows the classification of each construction site based on the percentage result obtained (Table 9) and the application of the classification proposed in Table 6.

Table 10 - Results of the sustainability classification of the construction sites.

WORKS	CLASSIFICATION OF SUSTAINABILITY ON CONSTRUCTION SITES
Work A	43% - Level 1 = Construction site with few sustainable practices.
Work B	44% - Level 1 = Construction site with few sustainable practices.
Work C	52% - Level 2 = Construction site with a good level of sustainable practices.
Work D	44% - Level 1 = Construction site with few sustainable practices.
Work E	51% - Level 2 = Construction site with a good level of sustainable practices.
Work F	59% - Level 2 = Construction site with a good level of sustainable practices.
Work F	36% - Level 1 = Construction site with few sustainable practices.
Work H	68% - Level 2 = Construction site with a good level of sustainable practices.
Work I	47% - Level 1 = Construction site with few sustainable practices.
Work J	54% - Level 2 = Construction site with a good level of sustainable practices.
Work L	44% - Level 1 = Construction site with few sustainable practices.
Work M	51% - Level 2 = Construction site with a good level of sustainable practices.
Work N	54% - Level 2 = Construction site with a good level of sustainable practices.
Work O	49% - Level 1 = Construction site with few sustainable practices.
Work P	43% - Level 1 = Construction site with few sustainable practices.

Source: The author (2016).

Based on the classification proposed in Table 6, it was possible to see that the construction sites were between level 1 and level 2 and no site was at level 0 or level 3.

Of the fifteen construction sites evaluated, eight were classified as "sites with few sustainable practices" and seven were classified as "sites with a good level of sustainable practices".

The most sustainable of these was site H, which obtained an average score of 68% on all the proposed topics, especially "Relationship with the environment" and "Waste generation".

On the other hand, site G was the least sustainable, with an average of 36% of the subjects covered.

The overall average of the fifteen construction sites evaluated was 49%, characterizing them as "sites with few sustainable practices".

Based on this information, it can be said that construction companies are applying environmental sustainability principles increasingly and gradually. However, it is clear that the increase in these sustainable actions is the result of legal requirements and not due to the companies' environmental awareness.

It can therefore be concluded that sustainable practices in the construction phase of buildings in Greater Vitória need to improve, as these are simple actions that, if carried out, can contribute to a better world.

5 CONCLUSION

The construction sector is moving towards industrialization, the introduction of sustainable practices and improvements in processes and products in a slow and unplanned way. The creation of laws, standards and resolutions are an excellent incentive for the development of the sector, but they need better parameters for implementation and, above all, enforcement.

Management tools, such as the application of a questionnaire on construction sites like the one carried out in this study, help in the application, control and evaluation of good practices, providing real *feedback on* how the company has been behaving in relation to sustainability.

The construction sector contributes directly to changing the physical space in which we live. It can alter the climate and produce waste and sediments that have a crucial impact on the environment.

Given these consequences, there is a need to create strategies to mitigate or cancel out these impacts.

As analyzed in this paper, the construction phase of a project is responsible for a large part of the environmental impacts, which is why construction sites deserve special attention, and all possible damage generated during construction must be identified and prevented.

In urban areas, where there is the greatest concentration of civil construction, it is necessary to take certain precautions to preserve the environment, to avoid sudden climatic changes, and to take care of the pollution generated, which degrades the environment and harms the neighborhood.

In this way, this study was motivated by curiosity and the need to find out what effective actions are being implemented by construction companies in Greater Vitória, ES, with regard to environmental sustainability practices in buildings, in order to remain competitive in the market in which they operate, as well as helping to minimize the impact generated by construction.

In this context, a qualitative survey of fifteen construction companies in Greater Vitória (ES) revealed that 53% of the sites had few sustainable practices and 47% had a good level of good practices. To this end, the results of most of the works assessed were unsatisfactory.

Although these are simple practices to carry out, their application is still low.

These are examples of actions that can be employed more frequently by construction companies:

- Planning to control the loss of materials in order to reduce the generation of rubble;
- Reuse of materials;
- Use of more efficient light bulbs and presence sensors to save electricity;
- Use of rainwater or artesian wells to save drinking water;

- Choosing material suppliers in locations closer to the development to reduce environmental impacts;

- A ban on smoking cigarettes on site to reduce air pollution in the workplace;

- Adoption of concrete mixer water decantation systems to avoid soil pollution;

- Adoption of the truck wheel-washing system to prevent pollution of the streets;

- Lectures on saving water and energy, environmental education and controlling waste generation to raise awareness of the work.

This study has shown how good practices are gradually being implemented by construction companies on their sites, since they have not yet reached satisfactory levels of sustainability.

REFERENCES

ABREU, W.G. **Sustainable Building Maintenance**: guidelines and practices in shopping centers. 2012. 150 f. Dissertation (Master's in Civil Engineering) - Post-Graduation in Civil Engineering, Universidade Federal Fluminense, Niterói, 2012.

AGOPYAN, V.; JOHN, V. M. **The challenge of sustainability in civil construction**. São Paulo: Blucher, 2012.

ARAÚJO, V. M. **Recommended practices for more sustainable construction site management**. 2009. 229 f. Dissertation (Master's in Civil Engineering) - Polytechnic School of the University of São Paulo, São Paulo, 2009.

ARAÚJO, V. M.; CARDOSO, F. F. **Survey of the state of the art: Construction sites**. 2007. Available at: <http://www.sindusconsp.com.br/img/meioambiente /20.pdf>. Accessed on: September 9, 2015.

BRAZILIAN ASSOCIATION OF TECHNICAL STANDARDS. **NBR ISO 14001**: Environmental management systems - Requirements with guidelines for use. Rio de Janeiro: 2004.

NBR 12284: Living areas on construction sites. Rio de Janeiro, 1991.

BIRBOJM, A. **Subsidies for decision-making regarding the choice of construction site elements**. 2001. 210 f. Dissertation (Master's in Civil Engineering) - Postgraduate Program in Civil and Urban Engineering, Polytechnic School of the University of São Paulo, São Paulo, 2001.

BITTENCOURT, M. **Avaliação dos Aspectos Ambientais em Canteiro de Obras.**2012. 246 f. Dissertation (Master's in Civil Engineering) - Graduate Program in Civil Engineering, Federal University of Santa Catarina, Florianópolis, 2012.

BLUMENSCHEIN, R. N. **Sustainability in the production chain of the construction industry**. 2004. 248 f. Thesis (Doctorate in Sustainable Development) - Center for Sustainable Development, University of Brasília, Brasília, 2004.

CARDOSO, R. S. **Sustainable construction sites**: a qualitative analysis of the impacts of implementation and operation. 2011. 66 f. Monograph (Graduation in Civil Engineering) - Federal University of Ceará, Fortaleza, 2011.

CESAR, D. L. et al. **Construction site design: evaluation of temporary facilities and physical material flows.** 2011. Available at: <http://www.iau.usp.br/ocs/index.php/sbqp2011/sbqp2011/paper/viewFile/268/214>. Accessed on: September 7, 2015.

NATIONAL ENVIRONMENTAL COUNCIL. **CONAMA Resolution No. 307**, of July 5, 2002. Available at: <http://www.mma.gov.br/port/conama/res/res02/ res30702.html>. Accessed on: September 6, 2015.

CREDÍDIO, F. **Sustainable buildings**: comfort and respect for the environment - Part 1. 2008. Available at:<http://www.institutofilantropia.org.br/component/ k2/item/2407sustainable_constructions_comfort_and_environment_respectp arte_1>. Accessed on: August 30, 2015.

DEGANI, C.M. **Environmental management systems in building construction companies**. 2003.

223 f. Dissertation (Master's in Civil Engineering) - Polytechnic School of the University of São Paulo, São Paulo, 2003.

DEGANI, C. M. **Sustainability management model for built facilities.** 2009. 210 f. Thesis (Doctorate in Civil Engineering) - Polytechnic School of the University of São Paulo, São Paulo, 2009.

DORSTHORST, B. J. H.; HENDRIKS, Ch. F. Re-use of construction and demolition waste in the Europe. **Proceedings of Symposium in Construction and Environment: Theory into Practice.** São Paulo: CIB; 2000.

VANZOLINI FOUNDATION. **AQUA - HQE process.** Available at: <http://vanzolini.org.br/conteudo-aqua.asp?cod_site=104&id_conteudo=1159>. Accessed on: October 10, 2015.

GEHLEN, J. **Building sustainability on construction sites:** a study in the Federal District. 2008. 158 f. Dissertation (Master's in Architecture and Urbanism) - CAPES Postgraduate Program, University of Brasília, Brasília, 2008.

GREEN BUILDING COUNCIL BRAZIL.**LEED New Construction.** Available at: <http://www.gbcbrasil.org.br/leed-new-construction.php>. Accessed on: October 10, 2015.

GUEDES, A. F. et al. **2014 World Cup - Mineirão Stadium and Sustainability Guidelines for the World's First Green Cup.**2011. Available at: < http://cbic.org.br/sites/default/files/Artigo%20copa%202014.pdf>. Accessed on: September 9, 2015.

INOJOSA, F. C. P. **Construction and demolition waste management:** CONAMA Resolution 307/2002 in the Federal District. 2010. 225 f. Dissertation (Master's in Civil Engineering) - Center for Sustainable Development, University of Brasília, Brasília, 2010.

JOHN, V. M. **Recycling waste in the construction industry**: a contribution to research and development methodology. 2000. 113 f. Thesis (Livre Docência) - Department of Civil Engineering, Polytechnic School of the University of São Paulo, São Paulo, 2000.

JOHN, V. M.; PRADO, R. T. A. **Selo casa azul**: Boas práticas para habitação mais sustentável. São Paulo: Páginas & Letras, 2010.

LAMBERTS, R.; TRIANA, M. A. **Survey of the state of the art:** Energy. São Paulo, 2007 Available at: <http://www.sindusconsp.com.br/img/meioambiente/ 16.pdf>. Accessed on: October 10, 2015.

LOBO, A. V. R. **Environmental sustainability assessment tool for hospital buildings in the metropolitan region of Curitiba.** 2010.   269 f.
Dissertation (Master's Degree in Civil Construction) - Postgraduate Program in Civil Construction, Federal University of Paraná, Curitiba, 2010.

LOTT, V. **LEED Mineirão Certification.** 2011. Available at: <http://www.copa2014.gov.br/sites/default/files/publicas/sobre-a-copa/camaras-tematicas/1-oficina-de-certificacao-e-gestao-sustentavel-das-arenas-copa-do-mundo-fifa-2014/arena-mineirao/ctmas-leed.pdf>. Accessed on: October 15, 2015.

MOURÃO, C.A.M.A.; NOVAES, M.V.; KEMMER, S.L. **Management of internal logistics flows in the construction industry** - the case of vertical **construction** projects in Fortaleza-CE. In:

BRAZILIAN SYMPOSIUM ON CONSTRUCTION MANAGEMENT AND ECONOMICS, 2009, João Pessoa, PB.

FILHO NETO, A. **Water as a building material.** 2013. Available at: <http://www.forumdaconstrucao.com.br/conteudo.php?a=43&Cod=625>. Accessed on: 08 Oct. 2015.

MARQUES NETO, J. da C. **Management of construction and demolition waste in Brazil.** São Carlos: Rima, 2005.

MINISTÉRIO DO TRABALHO E EMPREGO.**NR 18:** Condições e Meio Ambiente de Trabalho na Indústria da Construção. 2011. Disponível em: <http://portal.mte.gov.br /index.php/seguranca-e-saude-no-trabalho/2015-09-14-19-18-40/2015-09-14-19-23-50/2015-09-29-20-46-56>. Accessed on: September 2, 2015.

OLIVEIRA, J. A. da C. **Proposal for evaluating and classifying the environmental sustainability of construction sites:** ECO Methodology applied in the Federal District - DF. 2011.287 f. Thesis (Doctorate in Structures and Civil Construction) - Department of Civil and Environmental Engineering, University of Brasília, Brasília, 2011.

PESSARELO, R. G. **Water consumption on construction sites.** 2008. Available at: <http://www.revistasustentabilidade.com.br/artigos/consumo-de-agua-nos- canteiros>. Accessed on: August 7, 2015.

PINHEIRO, M. D. **Sustainable construction - myth or reality?** Lisbon, 2003. Available at: <https://fenix.tecnico.ulisboa.pt/downloadFile/3779571242058/Paper APEA_ConstrucaoSustentavel.pdf>. Accessed on: September 7, 2015.

PINTO, T. de P. **Gestão ambiental de resíduos da construção civil:** a experiência do Sinduscon - SP. São Paulo, 2005. Available at: <http://www.sindusconsp. com.br/downloads/prodserv/publicacoes/manual_residuos_solidos.pdf>. Accessed on: September 7, 2015.

PIRES, F. M. **Análise do comportamento sustentável das empresas do setor da construção civil da Grande Florianópolis.** 2008. 73 p. Monograph (Graduation in Economic Sciences) - Federal University of Santa Catarina, Florianópolis, 2008.

POMBO, F. R.; MAGRINI, A. **Panorama of application of the ISO 14001 standard in Brazil.** São Carlos, 2008. Available at:<http://www.scielo.br/scielo.php? script=sci_arttext&pid=S0104-530X2008000100002> Accessed on: Sep. 10, 2015.

RESENDE, F. **Atmospheric pollution by particulate matter emissions:** evaluation and control on building construction sites. 2007. 232 f. Dissertation (Master's in Civil Engineering) - Polytechnic School of the University of São Paulo, São Paulo, 2007.

ROCHA, E. G. de A. **Solid construction and demolition waste:** management, quantification and characterization - A case study in the Federal District. 2006. 174 f. Dissertation (Master's Degree in Structures and Civil Construction) - Department of Civil and Environmental Engineering, Faculty of Technology, University of Brasília, Brasília, 2006.

SCHENINI, P. C.; BAGNATI, A. M. Z.; CARDOSO, A. C. F. **Gestão de Resíduos da Construção Civil.** 2004. Available at < http://www.geodesia.ufsc.br/geodesia-online/arquivo/cobrac_2004/092.pdf>. Accessed on: September 12, 2015.

SILVA, V. G. **Sustainability assessment of Brazilian office buildings**: guidelines and methodological basis. 2003. 210 f. Thesis (Doctorate in Civil Engineering) - Department of Civil Construction Engineering, Polytechnic School, University of São Paulo. São Paulo, 2003.

TAVARES, D. K. P. A **method for evaluating waste management on construction sites**: Proposal based on construction companies in the city of Natal - RN. 2009. 160 f. Thesis (Master's in Production Engineering) - Production Engineering Program, Federal University of Rio Grande do Norte. Natal, 2009.

VERGARA, S. C. **Projetos e relatórios de pesquisa em administração**. 12. ed. São Paulo: Atlas, 2010.

VILHENA, J.M. **Diretrizes para a Sustentabilidade das Edificações.**   2007.
Available at: <http://www.revistas.usp.br/gestaodeprojetos/article/download/50905/   54986>.
Accessed on: September 9, 2015.

APPENDICES

APPENDIX A - Measures applied to construction sites included in the sustainability tools.

Chart 11 - Measures applied to construction sites included in the sustainability tools.

LEED	
Waste	Existence of bays with separation by waste class and correct waste disposal;
Energy	It doesn't;
Water	It doesn't;
Relationship with the environment	Allow the community around the site to communicate with the site, keep the site's surroundings clean;
Pollution	- Easy access for workers to public transport and the existence of bike racks on the construction site; - Use materials that have been manufactured close to the site (less than 500 km from the site); - Indoor tobacco smoke control; - Vehicle wheel cleaning device;
AQUA	
Waste	Existence of bays with separation by waste class and correct waste disposal;
Energy	- Awareness campaign to reduce electricity consumption on the construction site; - Use of low-energy light bulbs on the construction site;
Water	Implementation of water consumption control and reduction of drinking water consumption on the construction site;
Relationship with the environment	- Reduction of nuisance and pollution caused by the work (noise, visual, air pollution); -Circulation routes, parking spaces and suitable locations for delivering materials; - Maintaining the cleanliness of the construction site;
Pollution	- Vehicle wheel cleaning device; - Easy access for site workers to public transport and storage for bicycles; - Use materials that have been manufactured close to the site (within 300 km of the site);
BLUE HOUSE SEAL	
Waste	- Existence of bays with separation by waste class and correct waste disposal; - Reuse of forms and props;
Energy	- Awareness campaign to reduce electricity consumption on the construction site; - Ventilation and natural lighting in the temporary facilities of the construction site;

(continued)

BLUE HOUSE SEAL	
Water	Campaign to control water consumption and reduce drinking water consumption on the construction site;
Relationship with the environment	Achieving a good relationship with the community around the site through communication;
Pollution	- Encouraging workers to use bicycles; - Easy access for site workers to public transport; - Select low-impact materials; - Vehicle wheel cleaning device;

Source: The author (2015).

APPENDIX B - Questionnaire for assessing sustainability on

construction sites

in Greater Vitória

	Questionnaire for assessing sustainability on construction sites in Greater Vitória

Builder:
There are so many projects underway:
What type of work (commercial, residential):

QUESTIONS	YES	NO	NOTE
Do you have an ISO 14001 certificate?			
Do you have any sustainability tools certificates?			
Pollution			
Is there public transport near the site?			
Is an appropriate selection of materials and suppliers made, taking into account the reduction of environmental impacts and the emission of contributing gases?			
Does the construction site have bike racks?			
Do you have a system for washing vehicle wheels?			
Is smoking allowed on the construction site?			
Is the land located in an environmental preservation area?			
Are contractors included in environmental education programs for employees?			
Are stocks of oil, fuel and other flammable products identified?			
Are the rooms regularly humidified?			
Does it have a decanting system for water from washing the concrete mixer?			

(continued)

(continued)

QUESTIONS	YES	NO	NOTE
Water			
Does the site have a rainwater harvesting system or device?			
Does the site have an artesian well?			
Are there awareness talks on reducing water consumption on the construction site?			
Energy			
Are energy savings achieved through measures such as: prioritizing natural lighting and ventilation of the site's temporary facilities?			
Do you use compact fluorescent lamps?			
Does the construction site have sensors to trigger lighting in fast-passing environments?			
Are there any awareness talks on reducing energy consumption on site?			
Waste			
Do you implement solid waste management on the construction site?			
Is there selective waste collection on site?			
Are workers aware of waste management on site?			
Do you reuse materials?			
Do you control losses on construction sites?			
Is recyclable waste disposed of correctly?			
Do contractors receive any guidance on how to store the debris they generate?			

(continued)

QUESTIONS	YES	NO	NOTE
Relationship with the environment			
Are the surroundings (sidewalks and streets) clean?			
Are measures taken to control noise from equipment and activities in the vicinity?			
When materials leave the site in trucks, is there any prevention to avoid objects falling into the street?			
Are traffic problems prevented by controlling concreting times?			

Source: The author (2015).

ANNEXES

ANNEX A - Summary of categories, criteria and classification.

Table 12 - Summary of categories, criteria and classification.

CATEGORIES/CRITERIA		CLASSIFICATION		
1. URBAN QUALITY		**BRONZE**	**SILVER**	GOLD
1.1	Return Quality - Infrastructure	mandatory		
1.2	Quality of Surroundings - Impados	compulsory		
1.3	Improvements to the surroundings			
1.4	Recovering Degraded Areas			
1.5	Real Estate Rehabilitation			
2. DESIGN AND COMFORT				
2.1	**Landscaping**	mandatory		
2.2	Design flexibility			
2.3	Relationship with the Neighborhood			
2.4	Alternative Transport Solution			
2.5	Place for Selective Collection	ubrigalóriú		
2.6	Leisure Equipment Social and Sports	mandatory		
2.7	Thermal Performance - Fences	mandatory		
2.B	Thermal Performance - Orientation to Sun and Wind	compulsory		
2.9	Natural lighting in common areas			
2.10	Ventilation and Natural Lighting in Bathrooms			
2.11	Suitability for the Physical Conditions of the Site			
3. ENERGY EFFICIENCY				mandatory criteria - 12 free choice items
3.1	Low Energy Light Bulbs - Private Areas	mandatory for HIS - up to 3 6.M.	mandatory criteria + 6 **free** choice items	
3.2	Economizing Devices - Common Areas	mandatory		
3.3	Solar Heating System			
3.4	Gas Heating Systems			
3.5	Individualized Metering - Gas	mandatory		
3.6	Efficient elevators			
3.7	Efficient Appliances			
3.B	Alternative Energy Sources			
4. CONSERVATION OF MATERIAL RESOURCES				
4.1	Modular Coordination			
4.2	Quality of Materials and Components	úbifigalárlo		
4.3	Industrialized or prefabricated components			
4.4	Reusable forms and struts	oblige crio		
4.5	Construction and Demolition Waste (CDW) Management	mandatory		
4.6	Concrete with Optimized Dosage			
4,7	Blast Furnace Cement (CP1II) and Pozzolàmco (CP IV)			
4.8	Paving with **RCD**			
4.9	Easy façade maintenance			
4.1Q	Planted OR Certified Wood			

(continued)

(conclusion)

CATE GORY S/CRITERIA		BRONZE	CLASSIFICATION SILVER	GOLD
5. WATER MANAGEMENT				
5.1	Individualized Metering - Water	mandatory		
5.2	Economizing Devices - Discharge System	mandatory		
5.3	Ecano-m izers - Aerators			
5.4	Economizing Devices - Flow Regulator Register			
5.5	Use of rainwater			
5.6	Rainwater Retention			
5-7	Rainwater Infiltration			
5.0	Permeable areas	mandatory	mandatory criteria + 6 items of	mandatory criteria + 12 items
6. SOCIAL PRACTICES			free choice	free choice
6.1	Education for CDW Management	mandatory		
6.2	Environmental Education for Employees	mandatory		
6.3	Personal Development of Employees			
6-4	Professional Training for Employees			
6.5	Inclusion of local workers			
6.6	Community Participation in Project Development			
6.7	Orientation for Residents	mandatory		
6.6	Environmental Education for Residents			
6.9	Training for Enterprise Management			
6.10	Actions to Mitigate Social Risks			
6.11	Actions to Generate Employment and Income			

Source: John and Prado (2010).

Printed by Books on Demand GmbH, Norderstedt / Germany